My First Science Books

A CRABTREE SEEDLINGS BOOK

FOOD CHAIN IN A FOREST

Alan Walker

A Crabtree Seedlings Book

Crabtree Publishing

crabtreebooks.com

A forest is a **habitat**.

4

Animals find food and **shelter** there.

Grasshoppers munch on the grass and plants.

Grasshoppers are **herbivores**.

Mice nibble on seeds and insects.

Mice are **omnivores**.

Snakes slither through the forest. They hunt insects and mice.

Snakes are **carnivores**.

Night comes.
Owls catch their **prey**.

This is a forest food chain.

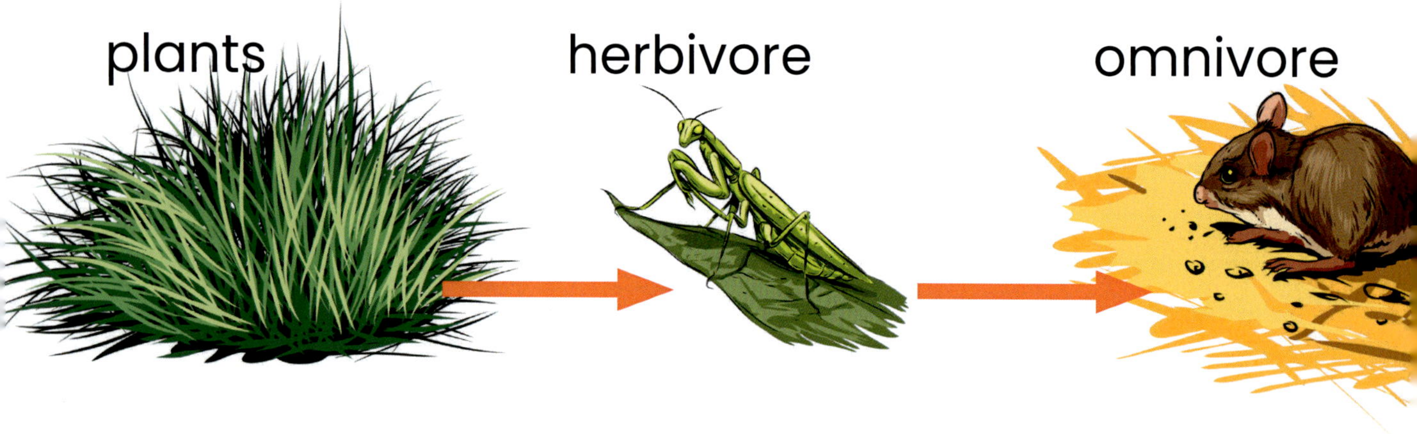

There is more than one food chain in a forest.

Read about other forest animals. Draw a different food chain.

Glossary

carnivores (KAR-nuh-vorz): Carnivores are animals that eat other animals.

habitat (HAB-uh-tat): A habitat is a place an animal naturally lives.

herbivores (HUR-buh-vorz): Herbivores are animals that eat only plants.

omnivores (OM-nuh-vorz): Omnivores are animals that eat both plants and other animals.

prey (PRAY): Prey are animals that are eaten by other animals.

shelter (SHEL-tur): A shelter is a place where an animal can live and hide from bad weather or danger.

Index

School-to-Home Support for Caregivers and Teachers

Crabtree Seedlings books help children grow by letting them practice reading. Here are a few guiding questions to help the reader with building his or her comprehension skills. Possible answers are included.

Before Reading

- **What do I think this book is about?** I think this book is about a forest food chain. It might tell us about the foods animals eat.
- **What do I want to learn about this topic?** I want to learn about the animals that are part of a forest food chain. Which four animals are on the cover?

During Reading

- **I wonder why...** I wonder why grasshoppers have such long antennae.
- **What have I learned so far?** I have learned that grasshoppers, mice, snakes, and owls are part of a forest food chain.

After Reading

- **What details did I learn about this topic?** I have learned that owls catch their prey at night.
- **Read the book again and look for the vocabulary words.** I see the word ***shelter*** on page 5 and the word ***carnivores*** on page 17. The other vocabulary words are found on pages 22 and 23.

Crabtree Publishing

crabtreebooks.com 800-387-7650

In Canada: We acknowledge the financial support of the Government of Canada through the Canada Book Fund for our publishing activities.

Hardcover 978-1-4271-3022-8
Paperback 978-1-4271-3033-4

Printed in Canada/042024/
CPC20240419

Published in Canada
Crabtree Publishing
616 Welland Avenue
St. Catharines, Ontario
L2M 5V6

Published in the United States
Crabtree Publishing
347 Fifth Avenue
Suite 1402-145
New York, NY 10016

Written by: Alan Walker

Photo Credits: Cover, Page 2-3 and 22 © Lapina Anna; page 4-5 and 23 Oleg Mayorov; cover, page 6-7 ans 22 © Gifu88; Page 8-9 © iava777; cover and page 14-15 © Bruce MacQueen; page 16-17 © Jason Patrick Ross; cover, page 18-19 and 22 © Michael Courtney; page 20-21 © NoPainNoGain, owl in food chain © Artur Balytskyi; page 23 burrow © Anitham Raju Yaragorla
All images from Shutterstock.com except:
Cover, Page 10-11 and 23 © Hans Hillewaert (image flopped) https://creativecommons.org/licenses/by-sa/3.0/deed.en ; Page 12-13 © Boschfoto https://creativecommons.org/licenses/by-sa/4.0/deed.en

Library and Archives Canada
Cataloguing in Publication
Available at the Library and Archives Canada

Library of Congress
Cataloging-in-Publication Data
Available at the Library of Congress